Count Me In

44 songs and rhymes about numbers

A&C Black · London

First published 1984 by
A & C Black (Publishers) Ltd
35 Bedford Row, London WC1R 4JH
© 1984 A & C Black (Publishers) Ltd.
Reprinted 1986, 1990

ISBN 0-7136-2622-4

Printed in Great Britain by Caligraving Ltd
Thetford, Norfolk

The activities and teaching suggestions were
devised by Beryl Boyd, Muriel Chester and
Margaret O'Shea

The musical arrangements are by Leonora Davies,
Brian Hunt, Peter Nickol, Jean Turnbull and
Sue Williams

The cover and illustrations are by Anni Axworthy

The songs and rhymes were chosen by Brian Hunt

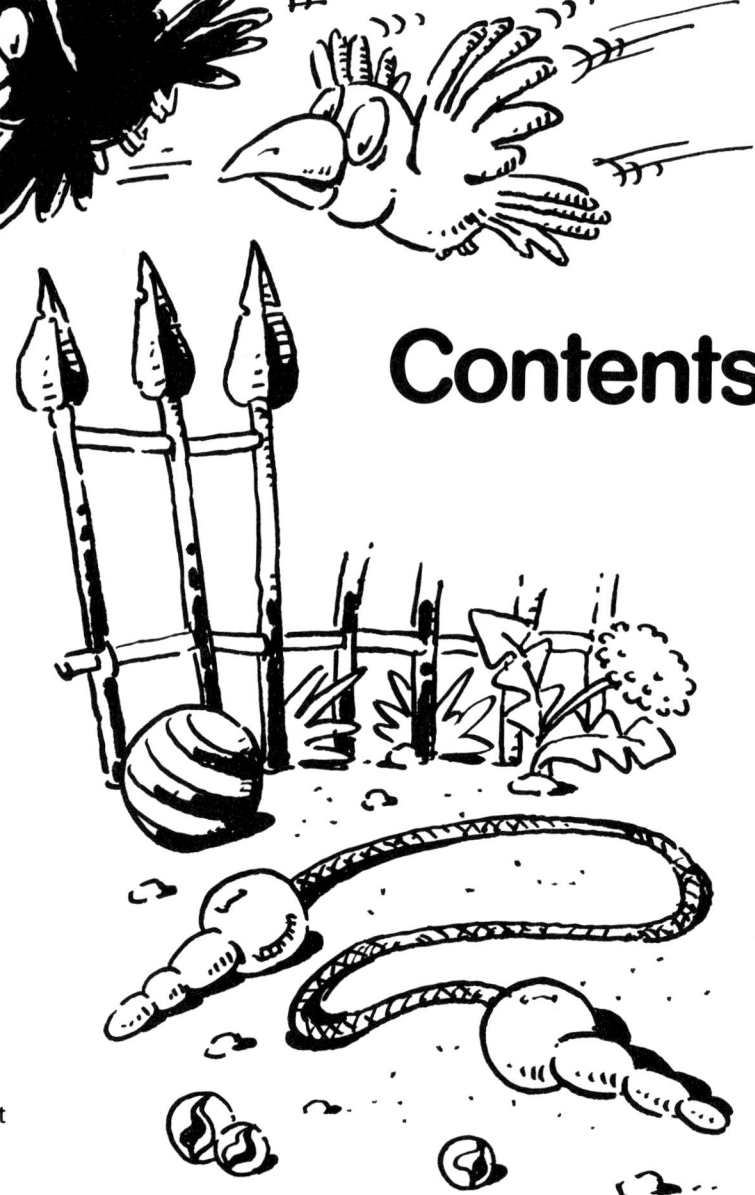

Contents

Introduction
Guitar chords

figures of fun

count up

count down

numbers at work

Introduction

This song book will be especially helpful with the teaching of basic number skills; there is an index of mathematical concepts at the back. Although the numerical aspect is significant to every item, 'Count Me In' is not a maths course set to music! You will be able to use the book on any occasion that calls for a good song or rhyme.

Guitar chords

X   string should not be sounded

'barre' – two or more strings held down by one finger

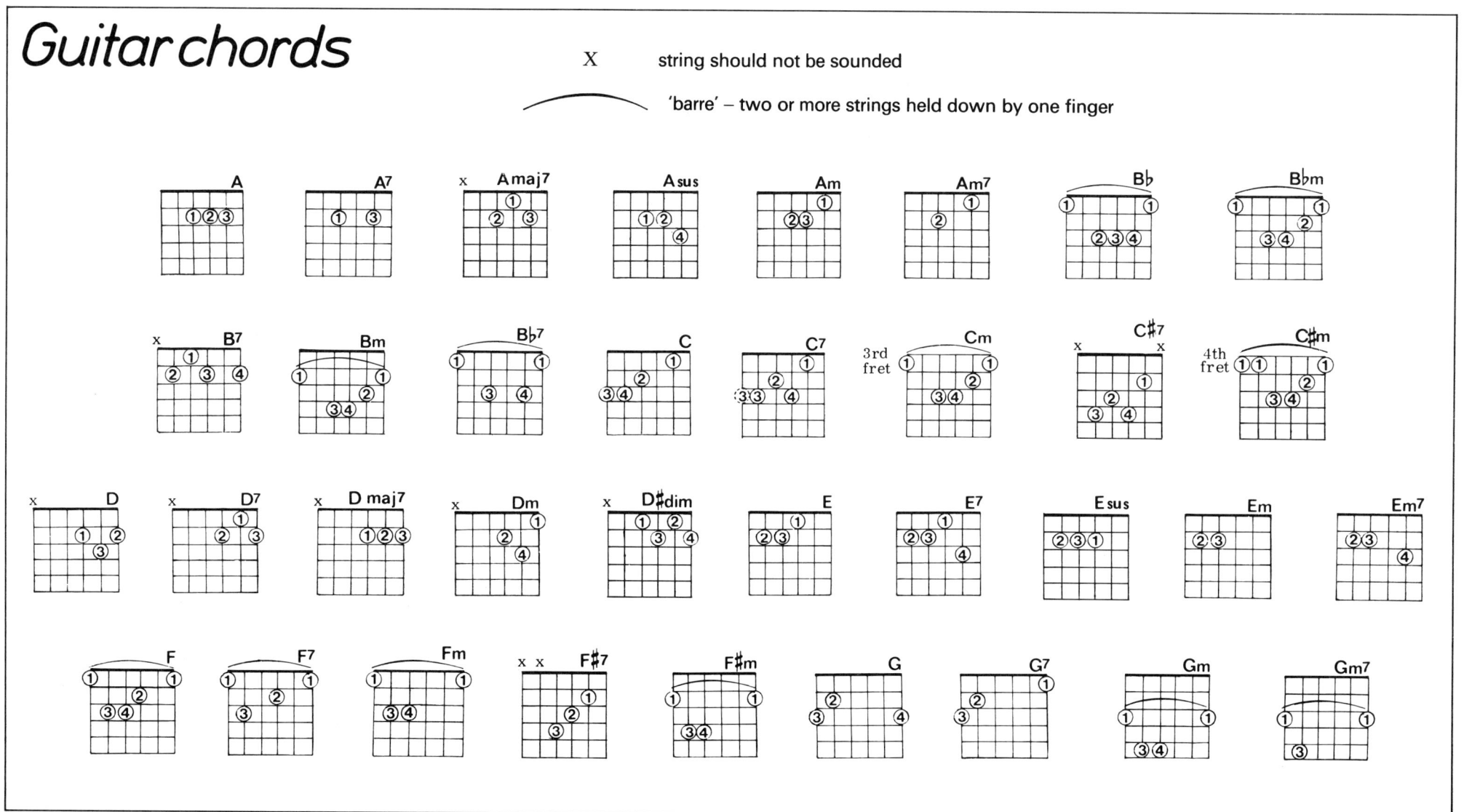

1 I have two ears

Gill Daniell

1   I have two ears to help me hear,
    I have two eyes to help me see,
    I have two lips to help me speak,
    Sing along with me.

    Two ears, one, two,
    Two eyes, one, two,
    Two lips, one, two,
    Sing along with me.

2   I have two legs to help me stand,
    I have two feet to help me dance,
    I have two hands so I can clap,
    Sing along with me.

    Two legs, one, two,
    Two feet, one, two,
    Two hands, one, two,
    I can count, can you?

> Try clapping in place of singing 'one, two'.
> The accompaniment may be omitted in the
> clapping bars.

2 White and brown

words: Clive Sansom
music: Jean Turnbull

My speckled hens,
    they've laid four –
Four little eggs
    in a nest of straw.
A white one for Margaret,
    a white one for Sue,
A white one for Peter,
    and a brown one for you!

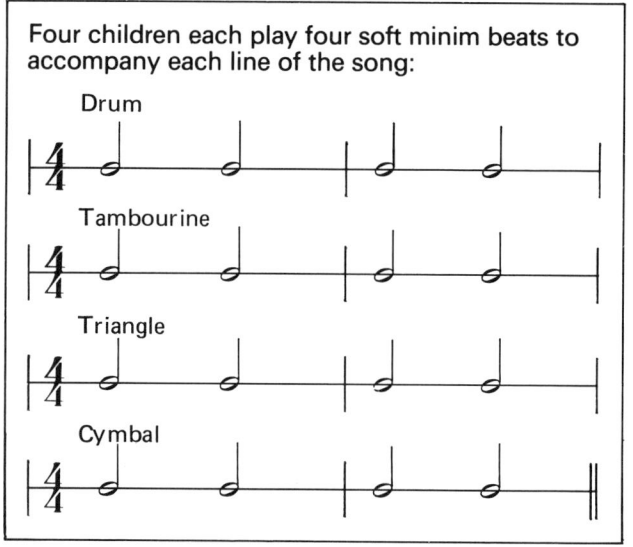

Four children each play four soft minim beats to accompany each line of the song:

Drum

Tambourine

Triangle

Cymbal

Place four 'eggs' – they can be made of plasticine – in a basket. Alternatively, cut egg-shapes with tabs from cardboard and slot them into a picture of a nest. As the song is sung, four chosen children each take an egg and put it in an egg-cup.

F (E*)   C (B7)

My speck - led hens,   they've laid   four –

C (B7)   F (E)

Four lit - tle eggs in a nest of straw. A

F (E)   C (B7)

white one for Mar-garet, a white one for Sue, A

C (B7)   F (E)   C (B7)   F (E)

white one for Pe -ter, and a brown one for you!

*capo on first fret may be used with alternative chords in brackets

3 Four seeds

words: traditional
music: Jon Betmead

Four seeds in a hole,
Four seeds in a hole –
One for the mouse,
One for the crow;
One to rot and one to grow!

4 Over in the meadow

traditional

1   Over in the meadow
    In the pond in the sun,
    Lived an old mother froggie
    And her little froggie one.
    'Hop' said the mother,
    'I hop' said the one
    So they hopped and were glad
    In the pond in the sun.

2   Over in the meadow
    In the nest in the tree,
    Lived an old mother birdie
    And her little birdies three.
    'Sing' said the mother,
    'We sing' said the three
    So they sang and were glad
    In their nest in the tree.

3   Over in the meadow
    In the sly little den,
    Lived an old mother spider
    And her little spiders ten.
    'Spin' said the mother,
    'We spin' said the ten
    So they span and were glad
    In the sly little den.

hopped and were glad In the pond in the sun.

5 Seven little pigs

Seven little pigs went to market,
One of them fell down,
One of them, he ran away,
And five got to town.

words: traditional
music: Margaret O'Shea

Se-ven lit-tle pigs went to mar-ket, One of them fell down,

One of them, he ran a-way, And five got to town.

Make the last line into a question:

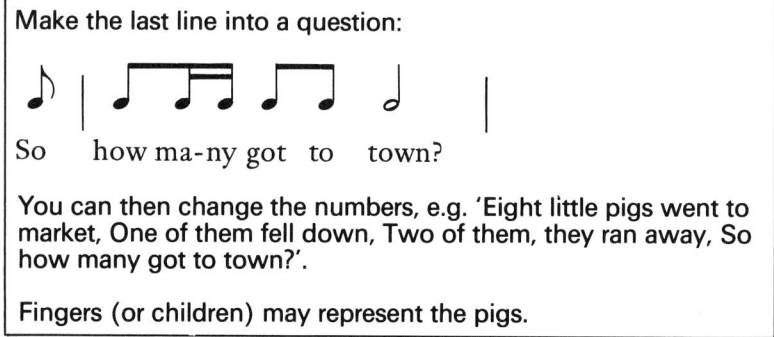

So how ma-ny got to town?

You can then change the numbers, e.g. 'Eight little pigs went to market, One of them fell down, Two of them, they ran away, So how many got to town?'.

Fingers (or children) may represent the pigs.

6 Me and you

traditional

I've got one head,
One nose too,
One mouth, one chin,
So have you.
I've got two eyes,
Two ears too,
Two arms, two legs,
And so have you.
I've got two hands,
Two thumbs too,
(Four fingers on each hand),
And so have you.

The children point to their own features while saying the rhyme. The last but one line might be omitted at first.

7 Five plump peas

words: traditional
music: Douglas Dimm

Five plump peas in a pea-pod pressed.
Five plump peas in a pea-pod pressed.
One grew, two grew,
And so did all the rest.
They grew and grew
And grew and grew
And grew and never stopped,
Till they grew so plump and portly
That the pea-pod – popped!

> Making a popping sound isn't difficult, but a growing sound may need some thought. A roll on a suspended cymbal or a rising note on a swanee whistle are two possibilities.

Five children could crouch in a line, growing with the song and eventually jumping in the air as the pod pops.

8 Only a boy named David

A S Arnott

Only a boy named David,
  only a rippling brook,
Only a boy named David,
  five little stones he took.
And one little stone went in the sling,
  and the sling went round and round.
One little stone went in the sling,
  and the sling went round and round.
Round and round and round and round,
  and round and round and round;
One little stone went up in the air
  and the giant came tumbling down.

*capo on first fret may be used with alternative chords in brackets

There are actions to this song. Try them every time you say these words:

only a boy

rippling brook

five little stones

one little stone

in the sling

round and round

up in the air

tumbling down

9 There were three furry cats

Gill Daniell

1 There were three furry cats
Purring loudly by the fire.
'Miaow, miaow, miaow'
Like a little kitten choir.

Can you count with me?
Come and let us see,
Can you count
One, two, three?

2 There were three little pigs
Grunting in their dirty sty,
'Oink, oink, oink'
To the people passing by.

3 There were three happy dogs
Chasing round in the sun.
'Woof, woof, woof
Aren't we all having fun!'

F (E*)

There were three fur – ry cats Pur – ring

Gm (F♯m)   C7 (B7)

loud – ly by the fire. 'Miaow, miaow, miaow' Like a

F (E)   C7 (B7)   F (E)

lit – tle kit – ten choir. Can you count with me? Come and

Gm (F♯m)   C7 (B7)   F (E)

let us see, Can you count One, two, three?

*capo on first fret may be used with alternative chords in brackets

10 There once was a sow

words: traditional
music: Mavis de Mierre

There once was a sow
Who had three piglets
And three piglets had she.
And the old sow always went
  'Umph',
And the piglets went
  'Wee, wee, wee'.

There once was a sow Who had three pig-lets

And three pig-lets had she.

And the old sow al – ways went

'Umph', And the pig-lets went 'Wee, wee, wee'.

11 Under a web

words: author unknown
music: Hugo Shortcombe

Under a web beside our gate
A spider hangs, his legs are eight.
Above him flies the busy bee,
Six black and furry legs has she.
A tabby cat goes leaping past,
Her four legs carry her so fast.
I've only two, that isn't many,
But Mr Worm, he hasn't any.

Make a drawing showing all the animals in the song and two children. Mark the faces of a cube 0, 2, 2, 4, 6 and 8 (to indicate numbers of legs). When the cube has been thrown, cover the appropriate picture (perhaps with a card showing the correct number of dots).

12 Chook-chook!

words: American rhyme
music: Jon Betmead

Chook, chook! chook-chook-chook!
– Good morning, Mrs Hen.
How many chickens have you got?
– Madam, I've got ten.
Four of them are yellow,
And four of them are brown,
And two of them are speckled red,
The nicest in the town.

Ask the children to draw ten chicks and colour them in according to the words of the song.

Change the numbers in the song, keeping a total of ten.

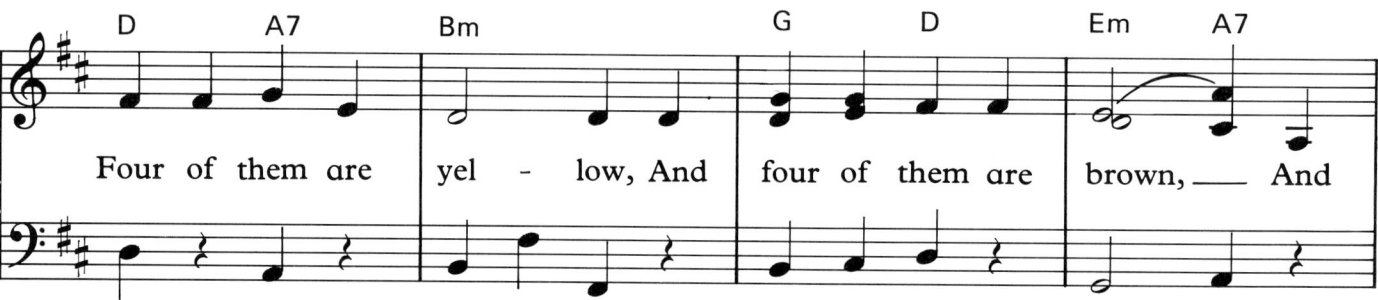

13 When I first to school was sent

Czech

1. When I first to school was sent
   I learnt how the numbers went:
   One and two, three and four,
   I know these and many more.

2. One, two, three, four, five and six,
   I can count them with my bricks:
   Dinner time soon begins –
   Baked potatoes in their skins.

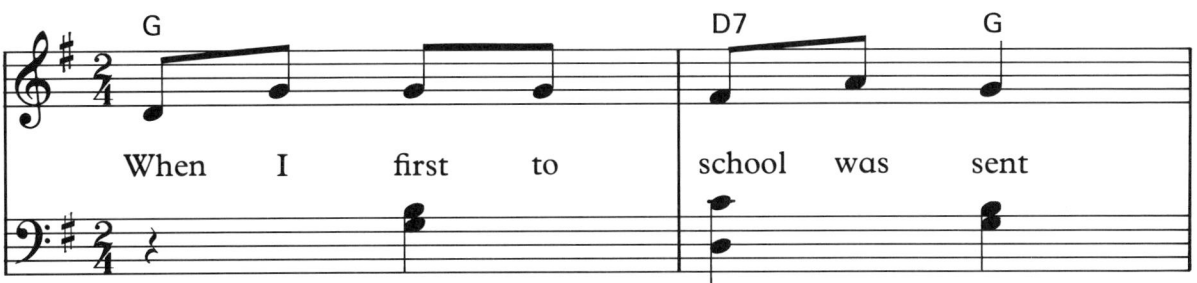

When I first to school was sent

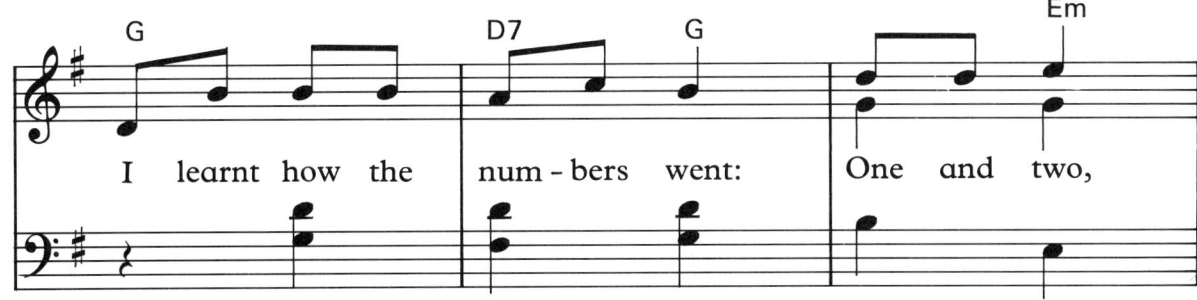

I learnt how the num-bers went: One and two,

three and four, I know these and ma-ny more.

14 Oliver Twist

Oliver Twist, can you do this?
If so, do so.
Number one, touch your tongue.
Number two, touch your shoe.
Number three, touch your knees.
Number four, touch the floor.
Number five, jump up high.
Number six, pick up sticks.
Number seven, fly to heaven.
Number eight, shut the gate.
Number nine, drink some wine.
Number ten, begin again.

This is a playground rhyme from Australia. A ball is bounced against a wall and the actions performed while it is in the air.

In the classroom, ten children could line up. Each child along the line performs one action while the class says the rhyme.

15 One, two, three a-learie

One, two, three a-learie,
Four, five, six a-learie,
Seven, eight, nine a-learie
Ten a-learie, postman.

This rhyme is from Britain. The ball is bounced on the ground as each number is said, then against a wall for 'a-learie'. On the word 'postman' the player spins round before catching the ball.

You may like to try a classroom version with clapping replacing the ball-bouncing.

One, two, three a - lear - ie, Four, five, six a - lear - ie,

Seven, eight, nine a - lear - ie Ten a - lear - ie, post - man.

16 The bee hive

words: traditional
music: Graham Westcott

Here is the bee hive,
    where are the bees?
Hidden away where nobody sees.
Soon they come creeping
    out of the hive,
One, two, three, four, five.

Here is one form of finger-play which can be used with this song:

the bee hive

where are the bees?

hidden away

they come creeping

one, two, three, four, five

Dm — Here is ___ the bee - hive,
A — where are ___ the bees?

A7 — Hid- den ___ a - way where
Dm — no - bod - y sees.

D7 — Soon they __ come creep - ing
Gm7 C7 — out of ____ the hive,

F — One,   two,
Dm A7 Dm — three, four, ___ five.

17 Band of angels

traditional spiritual

There was one, there were two,
There were three little angels,
There were four, there were five,
There were six little angels,
There were seven, there were eight,
There were nine little angels.
Ten little angels in that band.

Oh, wasn't that a band,
Christmas morning,
Christmas morning,
Christmas morning!
Oh wasn't that a band,
Christmas morning,
Christmas morning soon!

Choose ten 'angels'. They each decide which instrument they would like to play, and mime accordingly during the song. The first angel starts to play on the word 'one', the second on 'two' and so-on. They all play for the chorus. When will the music be loudest?

*capo on third fret may be used with alternative chords in brackets

18 Christmas candles

words: Clive Sansom
music: Graham Westcott

One . . . Two . . . Three –
Help me count the candles
On the Christmas tree.

Four . . . Five . . . Six –
I'll fix the candles,
You can light the wicks.

Seven . . . Eight . . . Nine –
In the winter twilight
How beautiful they shine.

Three – times – three:
What a lot of candles
On the Christmas tree!

One Two Three | Help me count the can - dles

On the Christmas tree. | Four Five Six | I'll fix the can - dles,

You can light the wicks. | Seven Eight Nine

In the win - ter twi - light How | beau - ti - ful they shine.

*capo on third fret will allow this song to be sung in key C as written

19 Log lore

One log won't burn
Two logs may;
Three logs must burn
And four will blaze away.

traditional

Chords: D, G, D, Bm7, E, A

Three times three: What a lot of can-dles On the Christ-mas tree!

20 Old John Braddle-um

traditional

1 Number one, number one,
  Now my song has just begun,
     With a rum-tum taddle-um,
     Old John Braddle-um,
     Hey what country folk we be.

2 Number two, number two,
  Some boots pinch, so I give a shoe,
     With a rum-tum . . .

3 Number three, number three,
  Some likes coffee and some likes tea,
     With a rum-tum . . .

4 Number four, number four,
  Some likes a gate, but I likes a door,
     With a rum-tum . . .

5 Number five, number five,
  Some folks die when they can't keep alive,
     With a rum-tum . . .

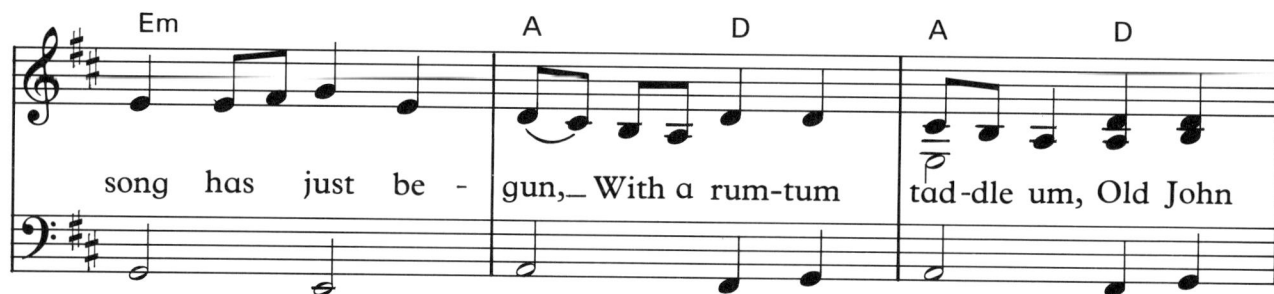

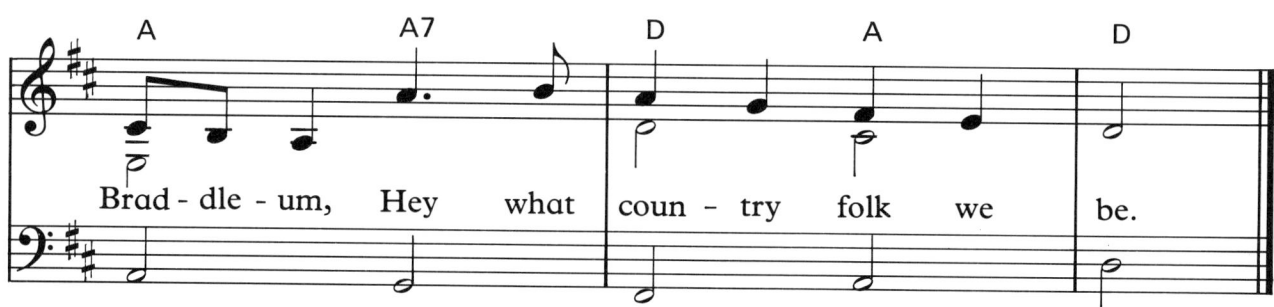

21 Counting steps

words: author unknown
music: Jean Turnbull

How many steps shall I have to take
To get from here to the door?
Please count how many steps I take
Walking across the floor.

    1, 2, 3, . . .

This song can be performed as a round

One child walks to the door while the rest count steps out loud. You can ask them to guess beforehand how many steps will be needed. Do big people take the same number of steps as little people? Change the starting point and the destination.

Sing just one phrase of the song and clap the beat. How many beats were there? Find out how many beats there are in the whole song.

22 Posting letters

The pillar box
Is fat and red
Its mouth is very wide.
I'm going to take some letters
And pop them all inside.

    1, 2, 3, . . .

A post box can be made from a cardboard box with a slit cut in it. One child posts letters while the others count out loud. Experiment with different letter sizes. Will they all fit through the mouth of the letter box?

author unknown

23 Come on board

words: author unknown
music: Graham Westcott

One is one and two is two.
I'm a space man. Who are you?

Three is three and four is four.
Listen to my space-ship's roar.

Five is five and six is six.
Come on board, we must be quick.

Seven is seven and eight is eight.
Would you like to be my mate?

Nine is nine and ten is ten.
You will not see earth again.

What instruments can be used to make 'space music'? Try to create a crescendo, at the end of which the rocket blasts off and gradually disappears from sight.

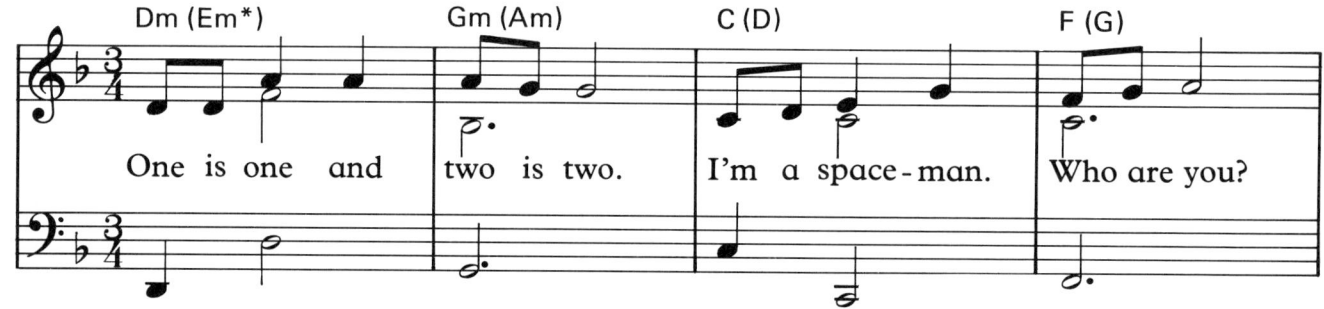

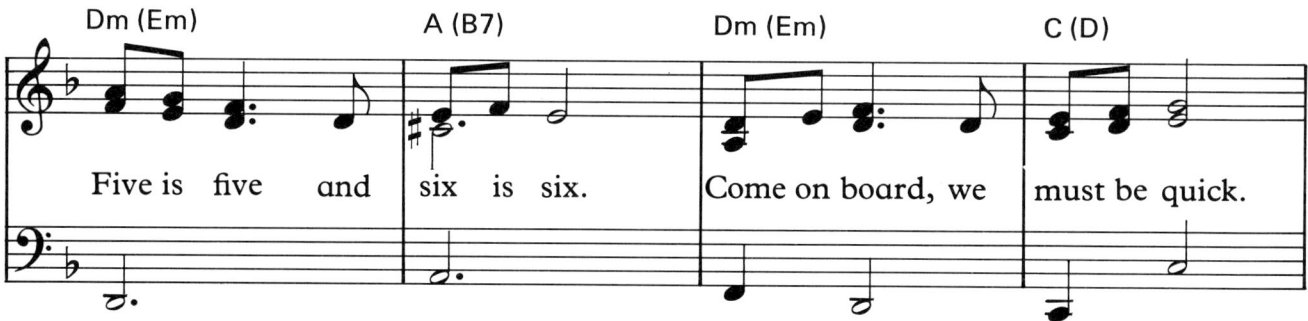

*guitarists may prefer to put this song into E minor

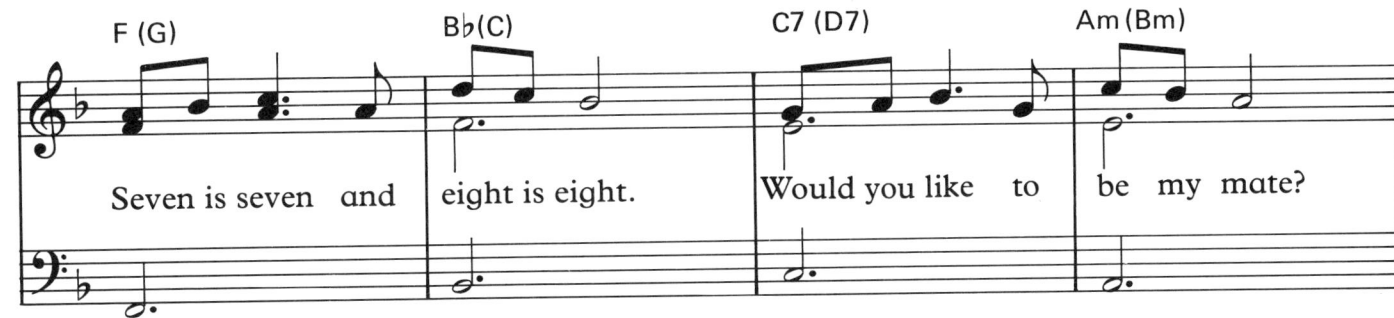

F (G)   B♭(C)   C7 (D7)   Am (Bm)

Seven is seven and | eight is eight. | Would you like to | be my mate?

Dm (Em)   Gm (Am)   F (G)   Am (Bm)   Dm (Em)

Nine is nine and | ten is ten. You will | not see earth a - | gain.

A space-ship game

Equipment:
- Playing boards showing a rocket marked 1 to 10 as a number-track.
- A spinner marked 1 to 10
- Sets of card or paper pieces marked with numbers of stars corresponding to the number track.

To play: Spin a number and place the correct star-piece on your rocket. The game continues until one rocket is ready for blast-off.

24 A chimney pot

words: traditional
music: Harriet Powell

I'm going to build a chimney pot
Very, very high.
I'll build it with my bricks,
And I'll make it touch the sky –
   1, 2, 3, . . .
Here's the wind and here's the rain
To knock my chimney down again.

If you have a lot of bricks you can carry on building and counting until the chimney topples. Otherwise, choose a number of bricks appropriate to the children's number development. A group of children could make the wind blow by flapping mats or papers. When the bricks have fallen down, ask 'Are there still the same number of bricks as there were in the chimney?'. Experiment with bricks of various sizes, finding out whether a taller chimney can be built by grading the sizes.

25 The apple-tree

words: Dorothy Williams
music: Graham Westcott

On the farmer's apple-tree,
Five red apples I can see.
Some for you, some for me –
Eat one apple from the tree.

On the farmer's apple-tree,
Four red apples I can see.
Some for you, some for me –
Eat one apple from the tree.

On the farmer's apple-tree,
Three red apples I can see.
Two for you, one for me –*
Eat one apple from the tree.

On the farmer's apple-tree,
Two red apples I can see.
One for you, one for me –
Eat one apple from the tree.

On the farmer's apple-tree,
One red apple I can see.
One for you, none for me –**
Eat one apple from the tree.

*or 'One for you, two for me'

**or 'None for you, one for me'
    or 'Half for you, half for me'

26 The banana seller

words: Sarah McNeill
music: Peter Hutchings

1 Little man standing in the market place,
   Shouting so loud he go all red in the face.
   Wanting the people his bananas to buy,
   And this is what he cry:
   One banana, two bananas, three bananas, four!
   Five bananas, six bananas, do you want some more?
   Seven, eight, or nine bananas! You can eat 'em raw!
   I can give you all my ten bananas! Tell me,
   What you waiting for?
   La-la-la, la-la-la! la-la-la, la-la-la!

2 Little man happy, and no wonder why!
   So many wanting his bananas to buy!
   He glad he got a lot bananas to sell,
   And all the time he yell:
   Ten bananas, nine bananas, eight bananas, see!
   I can sell you seven if you want to buy from me.
   Six, or five, or four bananas, I can sell you three.
   I can give you two or one bananas! Tell me,
   What you have for tea?
   La-la-la, la-la-la! La-la-la, la-la-la!

Try these patterns with the song:

Tambourine — What you have for tea?

Maracas — One ba-na-na for me

Wood block — Lit-tle man stand-ing there

27 Nine hairy monsters

words: Kaye Umansky
music: Douglas Dimm

Nine hairy monsters
Came to school today.
'Boo' said the teacher and
One ran away.
'Bong' went the bell
And the children ran to play
With eight hairy monsters
In the playground,
In the playground.

Eight hairy monsters . . .

Seven hairy monsters . . .

Six hairy monsters . . .

Five hairy monsters . . .

Four hairy monsters . . .

Three hairy monsters . . .

Two hairy monsters . . .

Nine hair-y mon-sters Came to school to - day.

'Boo' said the teach-er and One ran a - way.

'Bong' went the bell And the children ran to play With

eight hair-y mons-ters | In the play-ground, | In the play-ground.

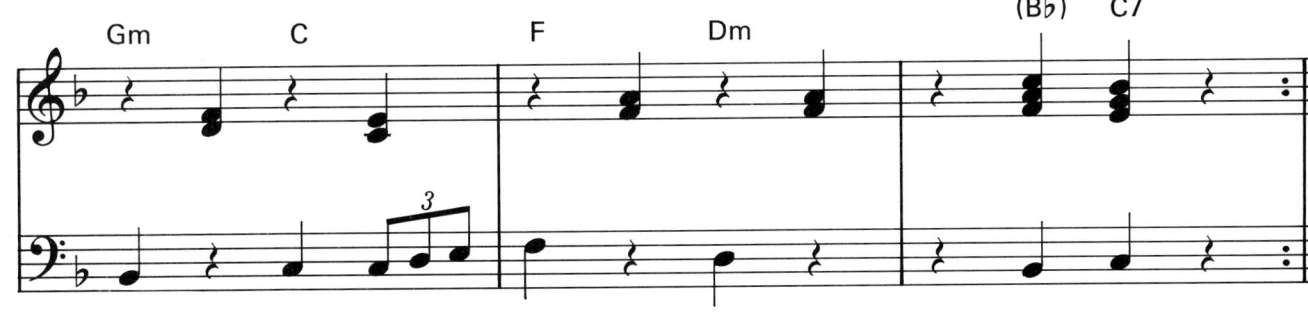

One hairy monster
Came to school today.
'Boo' said the teacher but
He didn't go away.
'Hurray' cried the children
As the teacher ran to play
With one hairy monster
In the playground,
In the playground.

Would the hairy monsters walk quickly or slowly? Would their hairy feet make a loud sound or a soft sound? It's easy to walk like a hairy monster to the beat of this song.

28 A pig tale

words: James Reeves
music: Alan True

Poor Jane Higgins,
She had five piggins,
And one got drowned in the Irish Sea.
Poor Jane Higgins,
She had four piggins,
And one flew over a sycamore tree.

Poor Jane Higgins,
She had three piggins,
And one was taken away for pork.
Poor Jane Higgins,
She had two piggins,
And one was sent to the Bishop of Cork.

Poor Jane Higgins,
She had one piggin,
And that was struck by a shower of hail.
Poor Jane Higgins,
She had no piggins,
And that is the end of my little pig tale.

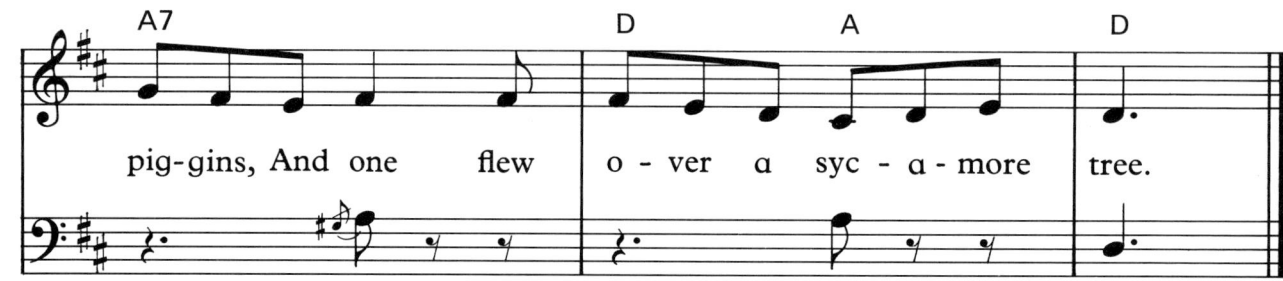

Have five piggins playing instruments, then four, then three, etc.

29 Falling leaves

words: Dorothy Williams
music: Judith Clingan

Five little leaves, so bright and gay,
Were dancing about on a tree one day.
The wind came blowing through the town –
    Phe – ew!
One little leaf came tumbling down.

Four little leaves . . .

Go on from here

Five lit – tle leaves, so bright and gay, Were
dan – cing a – bout on a tree one day. The
wind came blow – ing through the town – *Phe – ew!*
One lit – tle leaf came tumb – ling down.

30 Red balloons

words: Clive Sansom
music: Harriet Powell

Ten little red balloons,
    bobbing in a line;
One sailed away from them,
    and then there were nine.

Nine little red balloons,
    climbing up so straight:
One climbed crookedly,
    and then there were eight.

Eight little red balloons,
    floating up to heaven:
One came whirling down to earth,
    and then there were seven.

Seven little red balloons,
    getting in a fix:
One hit some power-wires,
    and then there were six.

Six little red balloons,
    trying to survive:
A small boy pricked one,
    and then there were five.

Five little red balloons,
    flying near the shore:
One touched the water,
    and then there were four.

Four little red balloons,
    sailing past a tree:
One got tangled in the leaves,
    and then there were three.

Three little red balloons,
    drifting past the zoo:
A tall giraffe swallowed one,
    and then there were two.

Two little red balloons,
    flying in the sun:
One went to find a star,
    and then there was one.

One little red balloon,
    his journey nearly done:
He went . . . POP! –
    and then there were none.

The rhythm of the third line is not the same for every verse.
You will need to work out how these lines fit the tune
before singing.

For each disappearance invent a sound which can be
played through bars nine and ten.

31 Ten green bottles

traditional
arranged by Leonora Davies

There were
Ten green bottles,
Hanging on the wall,
Ten green bottles,
Hanging on the wall,
And if one green bottle
Should accidentally fall –

There'd be
Nine green bottles,
Hanging on the wall . . .

Go on from here

The bottles are played by blowing across their necks. It might take a little practice, but the final effect can be most unusual. The children may need some help to find the right bottles, testing the pitch against a recorder or metallophone. You will probably have to add water to some bottles in order to produce the scale. One child plays each bottle in performance.

If bottles cannot be used, a treble recorder or alto metallophone could be a suitable substitute.

The drone (treble recorders) may be more effective if played on bass bars or on the piano two octaves lower than written. Be sure that the cymbal does not play before the word 'fall'.

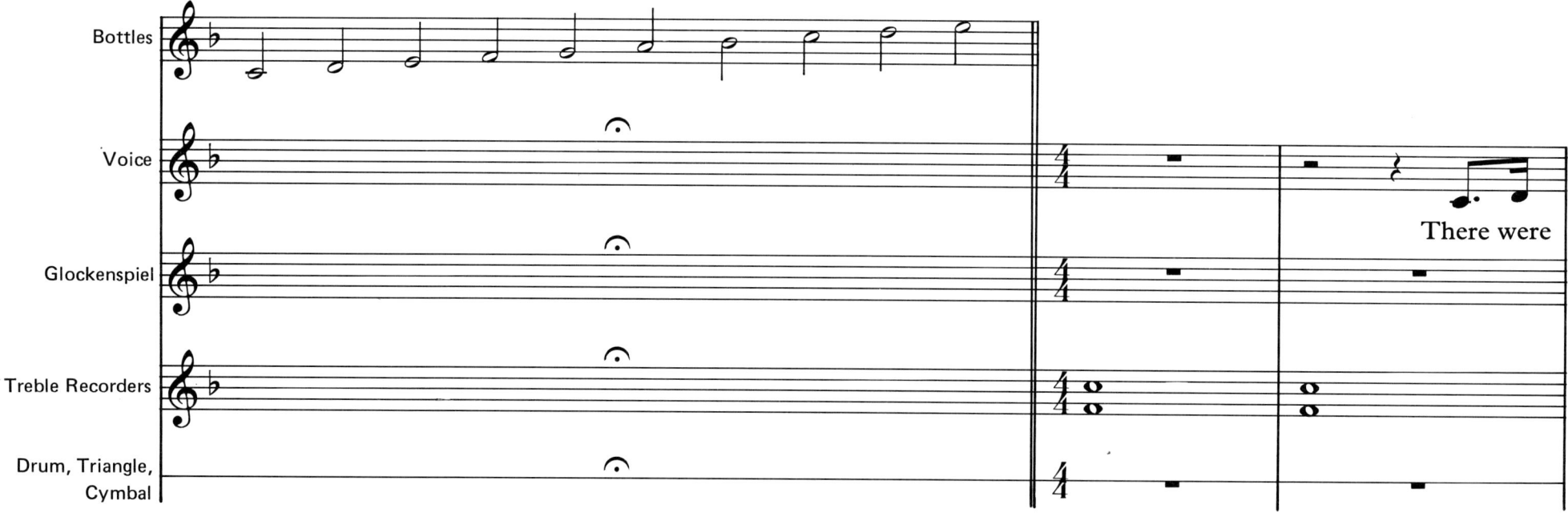

32 Six currant buns

words: traditional
music: Jean Turnbull

Six sticky buns in a baker's shop,
Big and brown with a currant on top.
A boy came along with a penny one day,
He paid one penny and took a bun away.

(repeat for 4, 3, 2 and 1)

No sticky buns in a baker's shop,
Big and brown with a currant on top.
A boy came along with a penny to pay.
'Sorry' said the baker,
'We have no buns left today.'

> This song can be acted, with one child as the baker who takes the pennies. Instead of 'a boy' sing the names of the individual children who buy buns.

*capo on third fret may be used with alternative chords in brackets

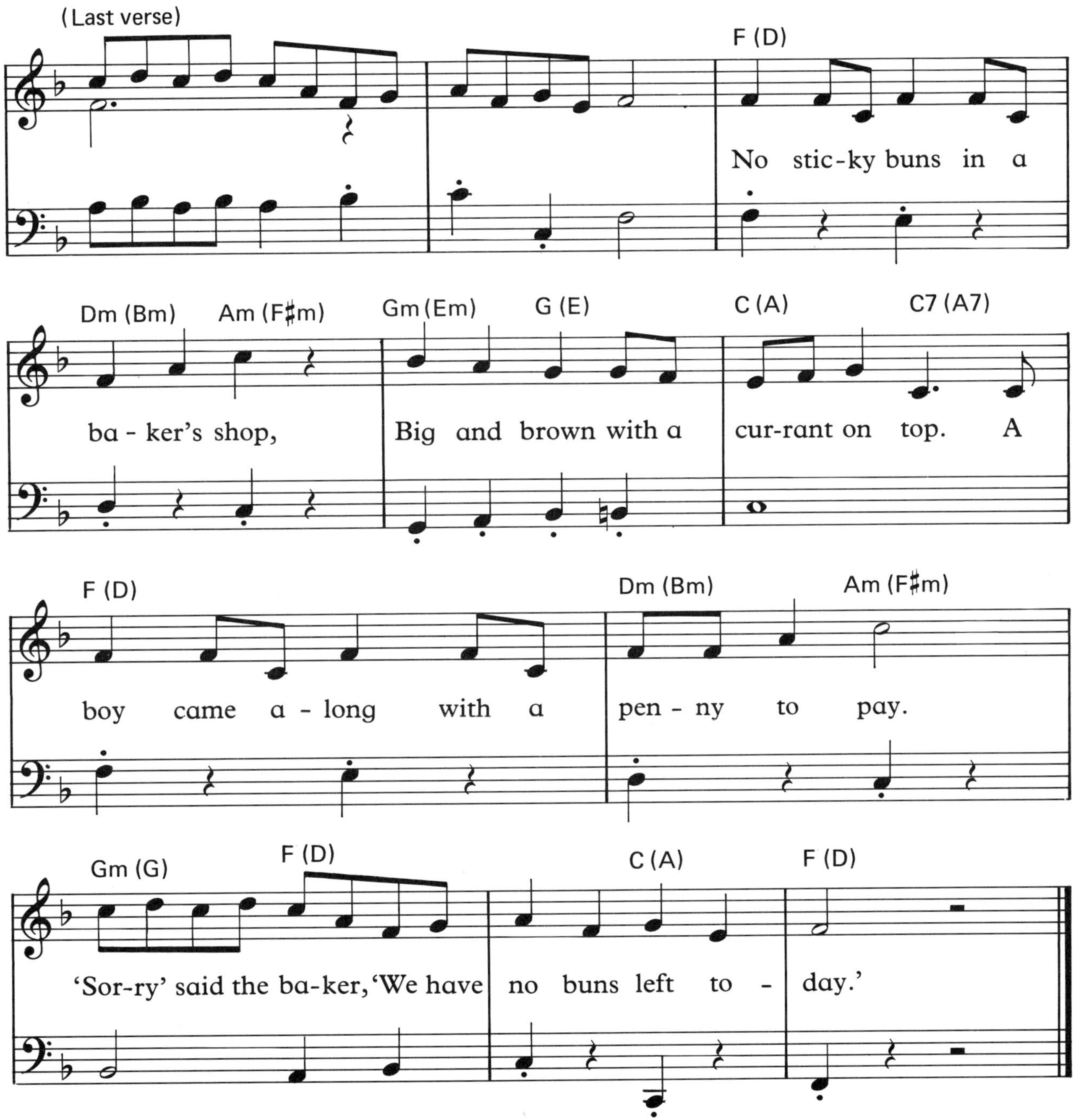
(Last verse)
F (D)
No stic-ky buns in a
Dm (Bm) Am (F♯m) Gm (Em) G (E) C (A) C7 (A7)
ba - ker's shop, Big and brown with a cur-rant on top. A
F (D) Dm (Bm) Am (F♯m)
boy came a - long with a pen - ny to pay.
Gm (G) F (D) C (A) F (D)
'Sor-ry' said the ba-ker, 'We have no buns left to - day.'

Baker's Shop
OPEN

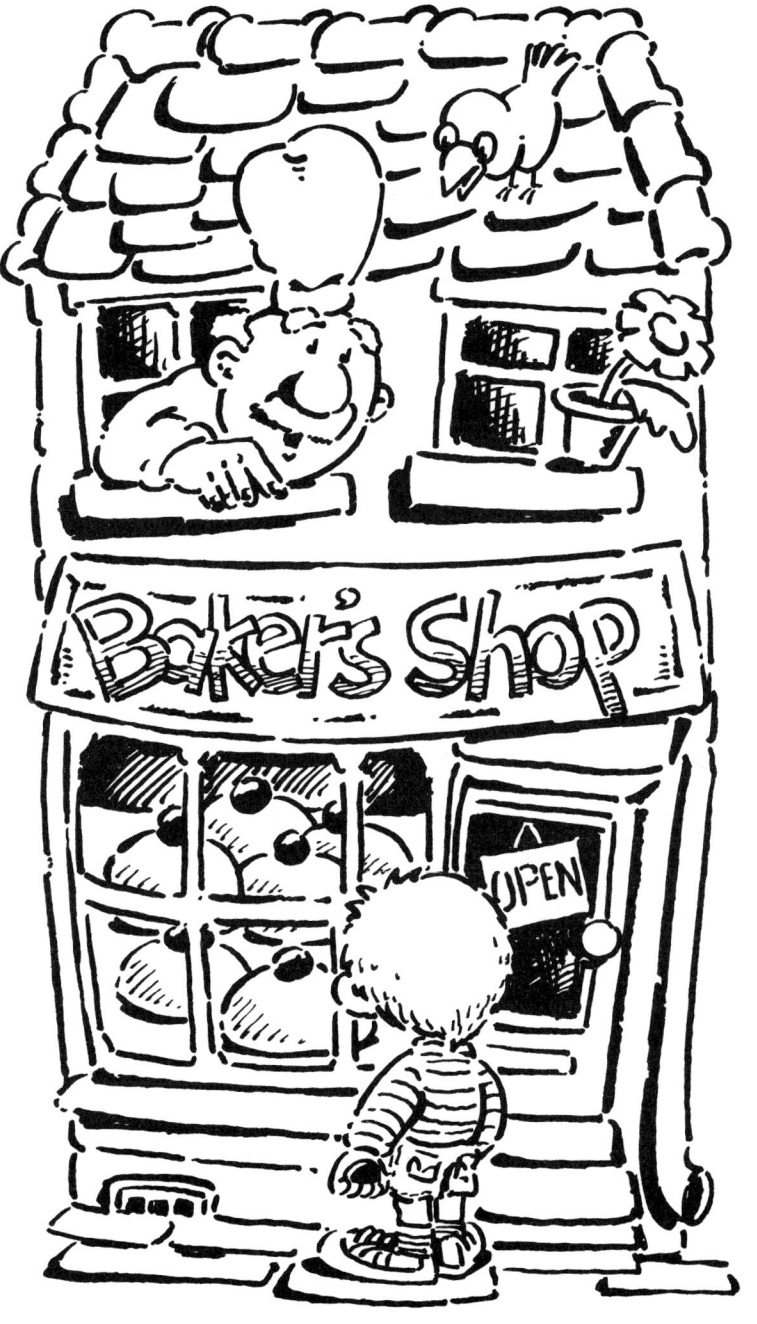

33 I've got sixpence

Box, Cox and Hall

1 I've got sixpence,
    Jolly, jolly sixpence.
    I've got sixpence
    To last me all my life.
    I've a penny to spend,
    I've a penny to lend
    And fourpence
    To take home to my wife.

    No cares have I to worry me,
    No clock upon the wall to hurry me.
    I'm as happy as a king,
    Believe me,
    As I go rolling home.

    Rolling home,
    Rolling home.
    As I go rolling home;
    I'm as happy as a king,
    Believe me,
    As I go rolling home.

2 I've got fourpence,
    Jolly, jolly fourpence . . .
    And twopence
    To take home to my wife . . .

3 I've got twopence,
    Jolly, jolly twopence . . .
    And nothing
    To take home to my wife . . .

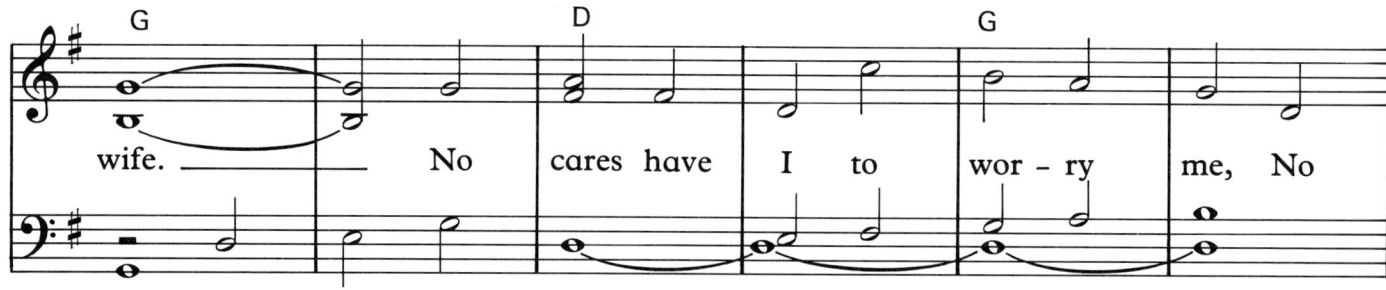

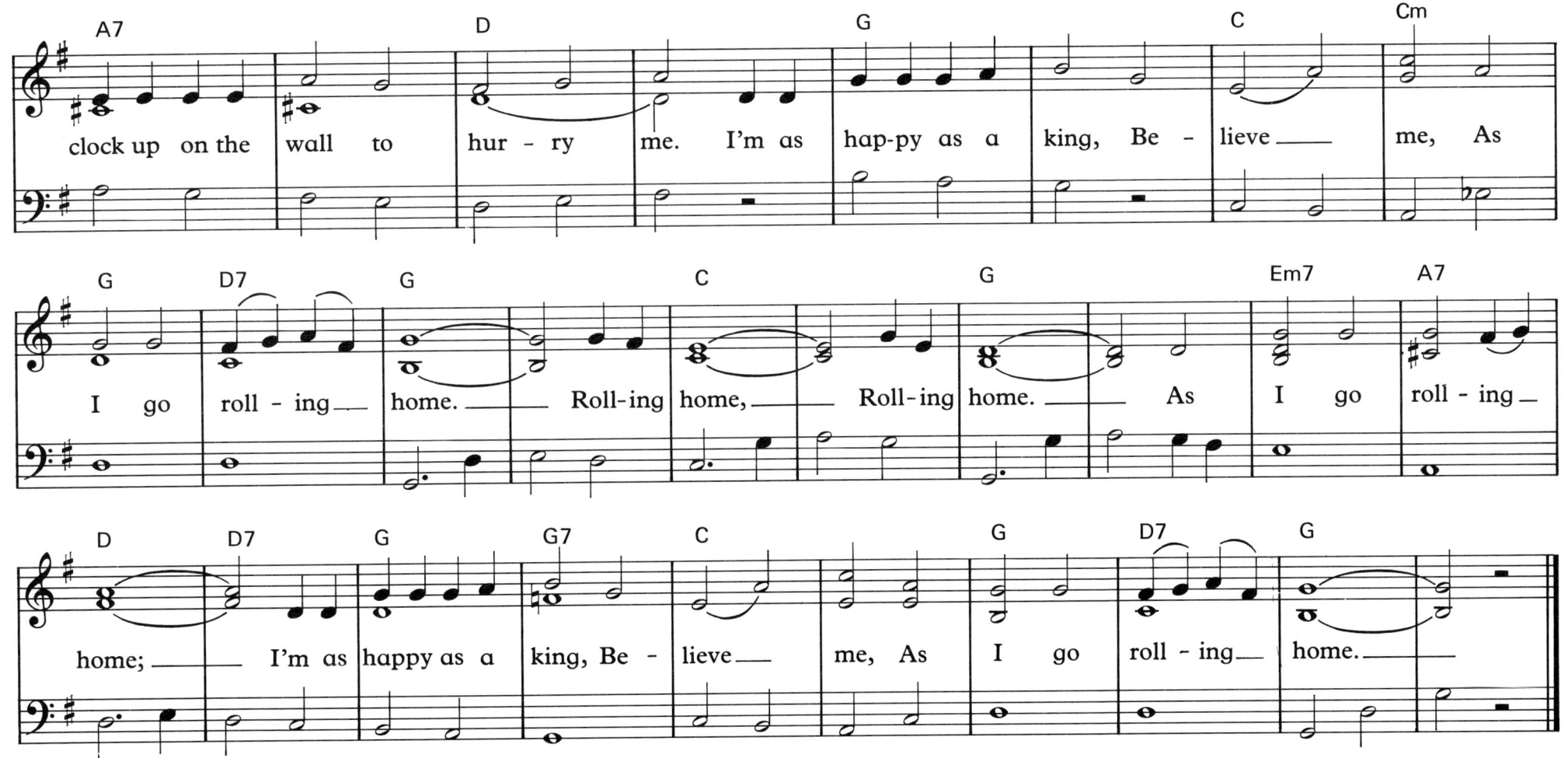

Many people today say 'two p' rather than 'tuppence', in which case the song must be about old money.

34 My first week at school

Kaye Umansky

1 On my first day at school
   My teacher gave to me
   One new pencil for writing.

2 On my second day at school
   My teacher gave to me
   Two green apples for biting.

3 On my third day at school
   My teacher gave to me
   Three old comics for reading.

4 On my fourth day at school
   My teacher gave to me
   Four fat rabbits for feeding.

5 On my fifth day at school
   My teacher gave to me
   Five fine conkers for weighing,
   On the sixth and seventh days
   I stayed at home
   And spent the weekend playing.

*capo on first fret may be used with alternative chords in brackets

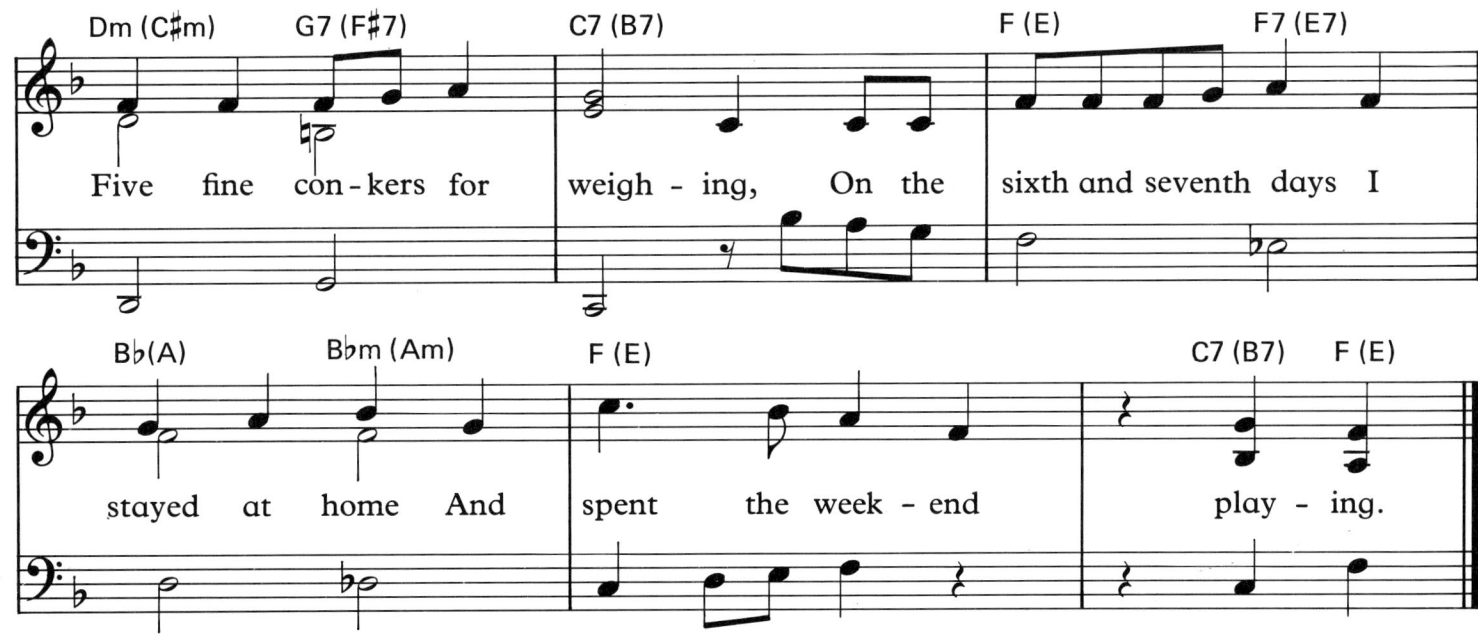

Five fine con-kers for weigh-ing, On the sixth and seventh days I stayed at home And spent the week-end play-ing.

Make a frieze:

35 Black Friars

words: Eleanor Farjeon
music: Leon Rosselson

Seven Black Friars,
Sitting back-to-back,
Fished from the bridge
For a pike or a jack.
The first caught a tiddler,
The second caught a crab,
The third caught a winkle,
The fourth caught a dab,
The fifth caught a tadpole,
The sixth caught an eel,
And the seventh one caught
An old cart-wheel!

Draw pictures of the different kinds of fish (including the cartwheel!) and attach the ordinal number names (first, second, etc). A jack is a young pike.

36 The centipede

Kaye Umansky

1 A centipede will certainly need
A hundred stripy socks,
But what'll he do when he wears them out
With climbing trees and rocks?
What'll he do when his socks wear through,
When all of his socks wear out?
He'll sit in a heap and start to weep
As his mother begins to shout:
– Here's what his mother will shout,
Whenever his socks wear out –

'I bought you ten, bought you twenty,
Bought you thirty, forty, fifty,
Bought you sixty, seventy, eighty, ninety,
Bought you a hundred socks!
So off to bed now, sonny,
Do you think I'm made of money?
Until I can afford to buy you more
You can keep your feet right off the floor!'

A cent - i - pede will cer-tain-ly need A

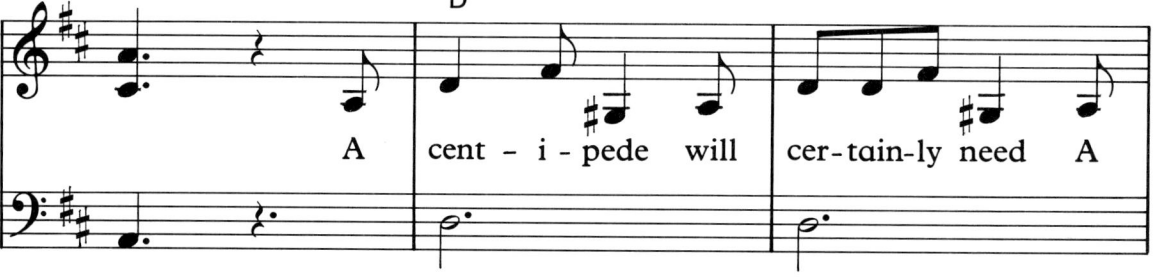

hun - dred strip - y socks, But what'll he do when he

D · · · Em · · · A

wears them out With | climb-ing trees and | rocks?

G · · · D · · · A7 · · · D

What-'ll he do when his | socks wear through, When

Em7 · · · A · · · D

all of his socks wear | out? He'll | sit in a heap and

Bm · · · A

start to weep As his | mo-ther be-gins to | shout:

TURN OVER

2 A centipede will certainly need
A hundred rubber boots,
But what'll he do when his boots wear through
With wriggling under roots?
What'll he do when his boots wear through
When all of his boots wear out?
He'll sit in a heap and start to weep
As his father begins to shout:
–Here's what his father will shout,
Whenever his boots wear out –

'I bought you ten, bought you twenty,
Bought you thirty, forty, fifty,
Bought you sixty, seventy, eighty, ninety,
Bought you a hundred boots!
So off to bed now, sonny,
Do you think I'm made of money?
Until I can afford to buy you more
You can keep your feet right off the floor!'

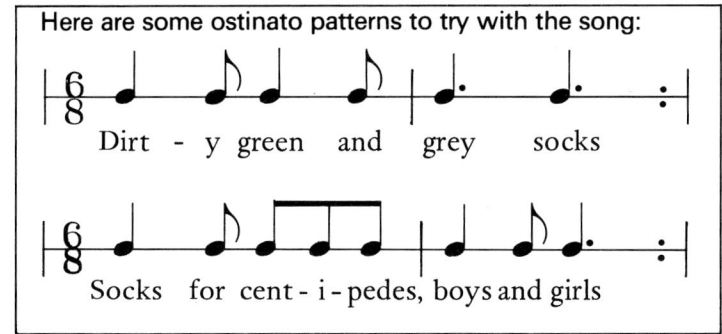

Here are some ostinato patterns to try with the song:

6/8 Dirt-y green and grey socks

6/8 Socks for cent-i-pedes, boys and girls

Make a project on socks – socks in the class, in football teams . . .
Activities could include ● sorting by colour, pattern or length
● counting pairs, matching pairs
● finding other things which come in pairs

37 Two in a boat

traditional American

Two in a boat and the waves run high
Two in a boat and the waves run high
Two in a boat and the waves run high
Get me a partner bye and bye.

Four in a boat . . .

Eight in a boat . . .

Two children sit on the floor, hands linked, and 'row' to the beat of the music. After 'bye and bye' each child chooses a new partner before the next verse begins.

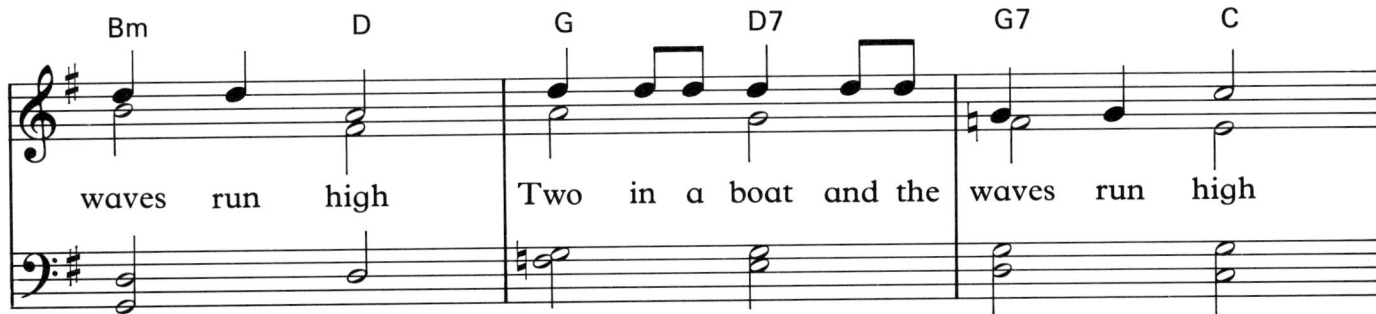

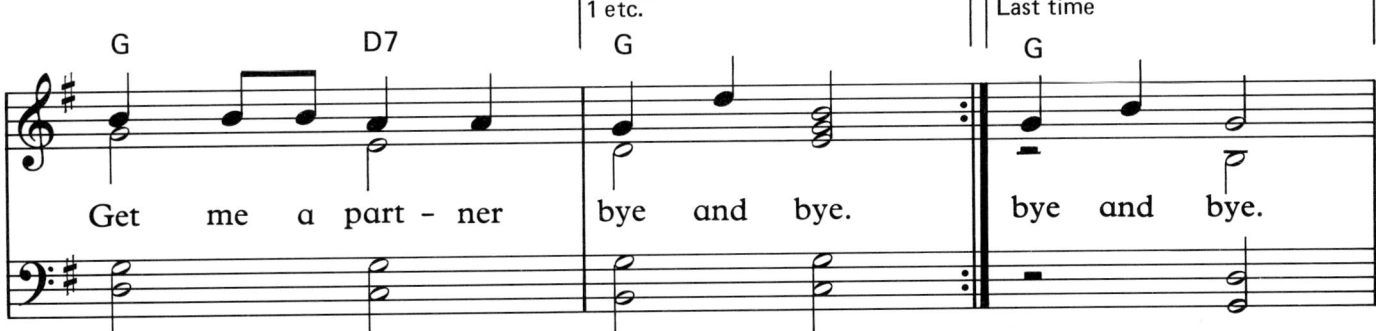

38 Twenty tomatoes

words: Kaye Umansky
music: Mavis de Mierre

I've got twenty tomatoes
'Cos I'm going to have a picnic,
Twenty should be plenty
For my friends and me.
Along came a policeman,
He arrested two tomatoes,
Now I've eighteen tomatoes
For a picnic tea.

I've got eighteen tomatoes . . .
Along came a burglar
And he stole two tomatoes,
Now I've sixteen tomatoes
For a picnic tea.

I've got sixteen tomatoes . . .
Along came a farmer
And he planted two tomatoes,
Now I've fourteen tomatoes
For a picnic tea.

I've got fourteen tomatoes . . .
Along came a cowboy
And he shot two tomatoes,
Now I've twelve red tomatoes
For a picnic tea.

I've got twelve red tomatoes . . .
Along came a giant
And he squashed two tomatoes,
Now I've ten red tomatoes
For a picnic tea.

I've got ten red tomatoes . . .
Along came a clown
And he juggled two tomatoes,
Now I've eight red tomatoes
For a picnic tea.

I've got eight red tomatoes . . .
Along came a witch
And 'Pufff!' went two tomatoes,
Now I've six red tomatoes
For a picnic tea.

I've got six red tomatoes . . .
Along came a goblin
And he gobbled two tomatoes,
Now I've four red tomatoes
For a picnic tea.

I've got four red tomatoes . . .
Along came a pirate
And his parrot liked tomatoes,
Now I've two red tomatoes
For a picnic tea.

I've got two red tomatoes
'Cos I'm going to have a picnic,
Two's not enough
To feed my friends and me.
I was getting hungry,
So I ate the two tomatoes,
Now there are no tomatoes
For my picnic tea.

39 Here comes the bus

Here comes the bus,
It's going to stop.
Hurry up children,
In you pop.
Four inside
And **six** on top.
How many altogether?

author unknown

> Change the numbers. If the rhyme is to be acted, a table can serve as a bus. The upper deck is on top of the table and the lower deck is underneath.

40 Crocodile!

A happy green crocodile
Eats **three** men.
Then he has the others – how many?
To make up to ten.

author unknown

> Change the numbers.

41 Three old men

words: Charles Henry Ross
music: Harriet Powell

Three old men sat a-thinking
For thirteen weeks and a day:
The first old man said nothing,
And the second old man said less,
So the third old man walked away.

> You can introduce ordinal numbers
> (first, second, etc) with this song. Be
> sure that the joke about less than
> nothing doesn't confuse.

Three old men   sat a-think-ing

For thir-teen weeks and a   day:_____   The   first   old   man   said

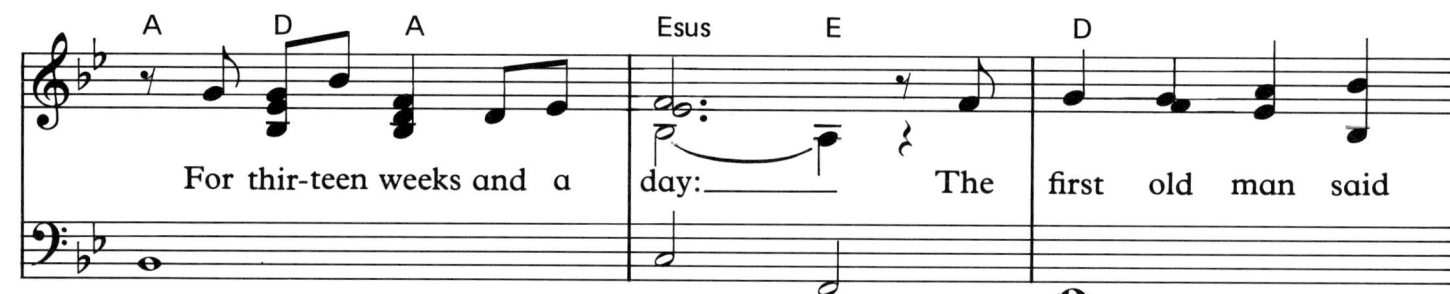

no-thing,   And the   se-cond old man said   less,   So the

third   old   man   walked a - way.   The   first   old   man   said

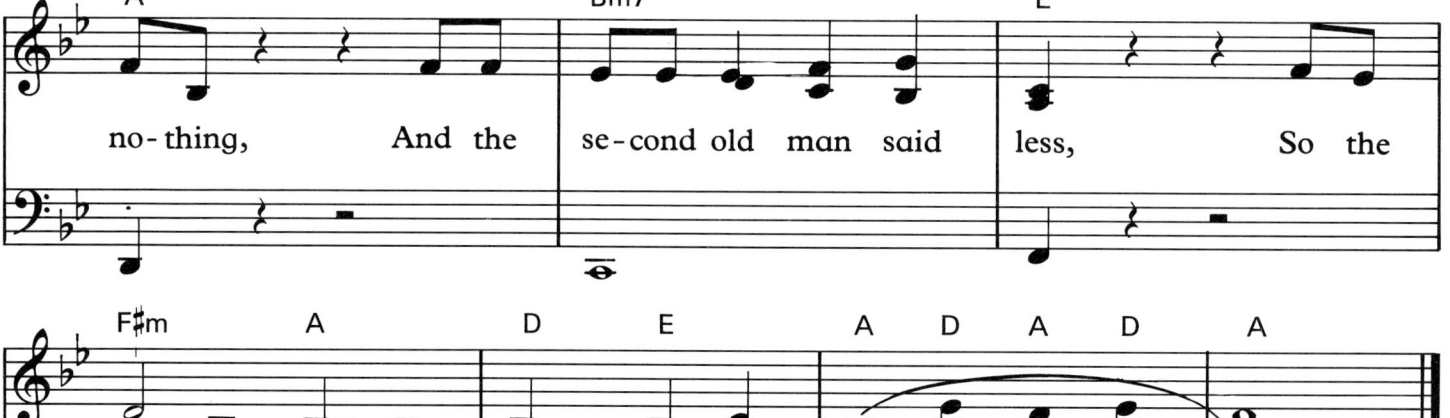

*capo on first fret will allow this song to be sung in key B flat as written

42 The King of Roly Poly

Kaye Umansky

The King of Roly Poly
He's got seven fine sons –
Seven fine sons so tall.
Three like jelly sandwiches
And three like Chelsea buns,
But the little one likes custard,
Yes, custard best of all.

The Queen of Lemon Barley
She's got five fine daughters –
Five fine daughters tall.
Two like fizzy cherryade
And two like soda water,
But the little one likes lime juice,
Yes, lime juice best of all.

The Duke of Sticky Toffee
He's got three fine sons –
Three fine sons so tall.
Two like eating candy bars
(The strawberry flavoured ones),
But the little one likes sherbert,
Yes, sherbert best of all.

The Duchess of The Dairy
She's got one fine baby –
One fine baby small.
He'll get to like a lot of things
When he gets older, maybe,
But the milk is what he likes now,
Yes, likes it best of all.

43 The appleman

words: Helen Clyde
music: Margaret O'Shea

Well, well, well,
Which of you can tell?
How many apples
Did the old man sell?
One on Monday,
Two on Tuesday,
Three on Wednesday,
Four on Thursday,
Five on Friday,
Six on Saturday,
Well, well, well,
Now can you tell? –
How many apples did the old man sell?

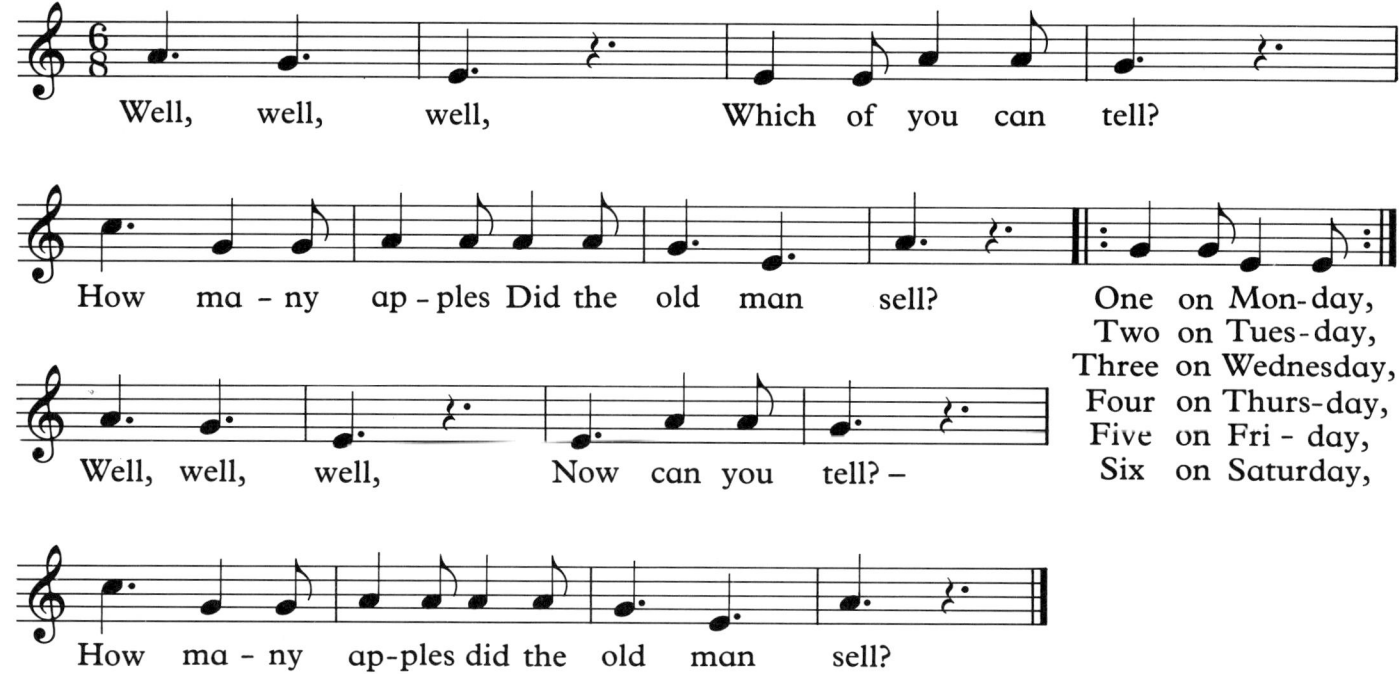

Well, well, well, Which of you can tell?

How many apples Did the old man sell?

One on Mon-day,
Two on Tues-day,
Three on Wednesday,
Four on Thurs-day,
Five on Fri - day,
Six on Saturday,

Well, well, well, Now can you tell? –

How many apples did the old man sell?

The apples can be shown on a triangular tree:

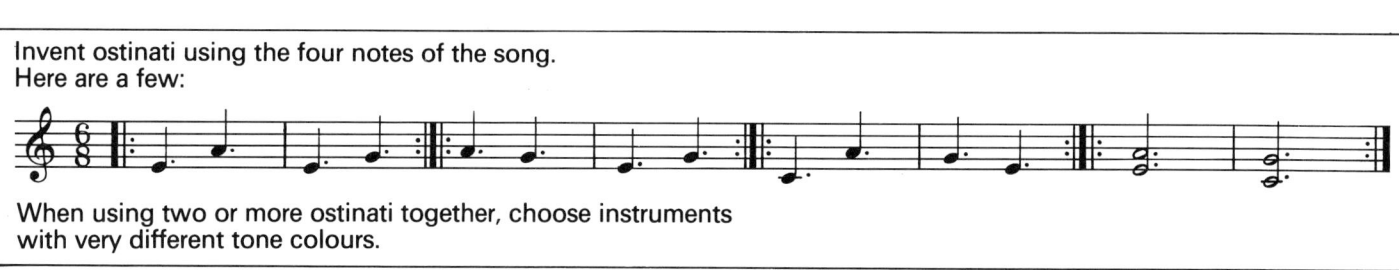

Invent ostinati using the four notes of the song.
Here are a few:

When using two or more ostinati together, choose instruments
with very different tone colours.

Vary the numbers, e.g. sing 'two on Monday,
four on Friday...'. At the end of the song the
children give the answer.

44 Noah's birthday

Kaye Umansky

This is a song for virtuoso mathematicians! If you want an easier version, change the line marked * to 'All in a row, all in a row'.

Rat - a - tat - tat on the door of the ark The post is brought by a whale and a shark; Seven par - cels all in a row, To - day is No - ah's birth - day.

TURN OVER

Rat-a-tat-tat on the door of the ark
The post is brought by a whale and a shark;
Seven parcels all in a row,
Today is Noah's birthday.

1 Open the first, what have we here?
A lovely pair of elephant ears.
Thank you, Elephant, thank you dear,
It's what I've always wanted.

    Seven parcels all in a row,
    One is open, six to go.★
    Seven parcels all in a row,
    Today is Noah's birthday.

2 Open the second, what have we here?
The wallaby sent some ginger beer.
Thank you Wallaby, thank you dear,
It's what I've always wanted.

    Seven parcels all in a row,
    Two are open, five to go . . .

3 Open the third, what have we here?
A tiddly jar of crocodile tears.
Thank you Crocodile, thank you dear,
It's what I've always wanted.

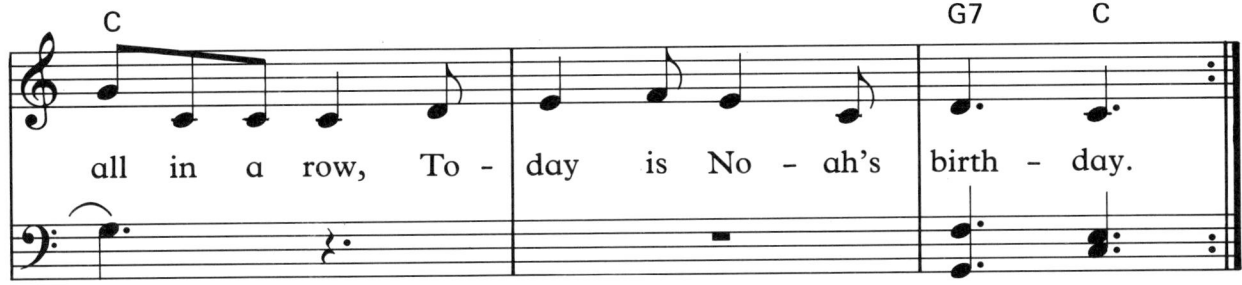

all in a row, To - day is No - ah's birth - day.

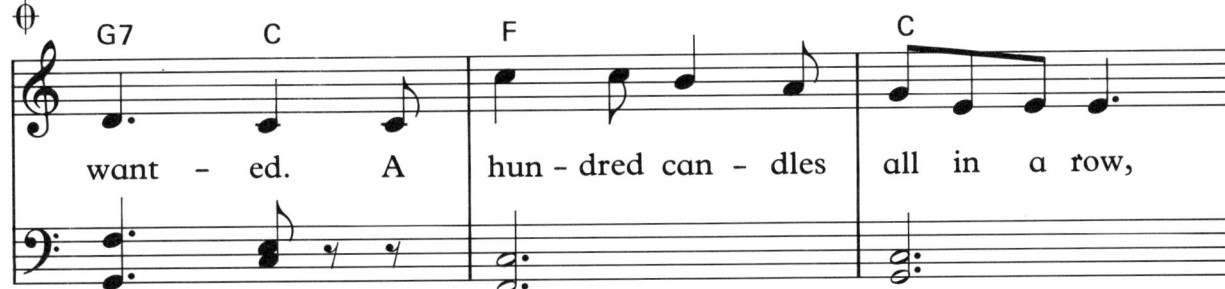

want - ed. A hun - dred can - dles all in a row,

Give 'em a blow, give 'em a blow; A hun - dred can - dles

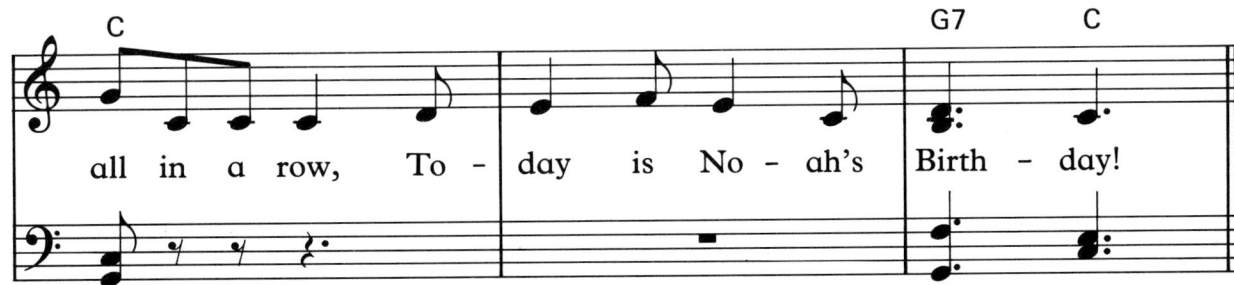

all in a row, To - day is No - ah's Birth - day!

4 Open the fourth, what have we here?
A pair of antlers from the deer.
Thank you Deer, oh thank you dear,
It's what I've always wanted.

5 Open the fifth, what have we here?
The monkey sent a chandelier.
Thank you Monkey, thank you dear,
It's what I've always wanted.

6 Open the sixth, what have we here?
It's much to small to see very clear.
Thank you Ant, oh thank you dear,
I'm sure it's what I wanted.

7 Open the last, what can it be?
His sons have baked a cake for tea!
Thank you fellas, it's luv-er-ly,
And what I've always wanted.

A hundred candles all in a row,
Give 'em a blow, give 'em a blow;
A hundred candles all in a row,
Today is Noah's Birthday!

Maths index

Where the particular mathematical concept is not central, the number of a song or rhyme is in brackets.

Acknowledgements

The following copyright owners have kindly granted their permission
for the inclusion of these items:

1 I have two ears, 9 There were three furry cats
'Words and music reprinted by permission of Mercury Music Company Ltd.
Songs from the Granada Television series 1...2..3.. Go!.'

2 White and brown
Ruth Sansom (words); Jean Turnbull (music)

3 Four seeds, 12 Chook-chook!
Jon Betmead (music)

5 Seven little pigs
Margaret O'Shea (music)

8 Only a boy named David
'© 1931 Salvationist Publishing and Supplies Ltd, London' (words and music)

10 There once was a sow, 38 Twenty tomatoes
Mavis de Mierre (music)

13 When I first to school was sent
'© 1975 Schott & Co Ltd' (words and music)

14 Oliver Twist
Heinemann Educational Australia Ltd, from *Cinderella Dressed in Yella*,
Australian children's play-rhymes collected by Ian Turner

16 The bee hive
Graham Westcott (music)

18 Christmas Candles
Ruth Sansom (words); Graham Westcott (music)

21 Counting steps
Jean Turnbull (music)

23 Come on board
Graham Westcott (music)

24 A chimney pot
Harriet Powell (music)

25 The apple-tree
Clarendon Press, from *A Realistic Approach to Number Teaching*
by Dorothy Williams (words); Graham Westcott (music)

26 The banana seller
BBC Publications, from *Maths Songbook* (words and music)

28 A pig tale
Oxford University Press, from *The Blackbird in the Lilac*
by James Reeves (words); Alan True (music)

29 Falling leaves
Clarendon Press, from *A Realistic Approach to Number Teaching*
by Dorothy Williams (words); Judith Clingan (music)

30 Red balloons
Ruth Sansom (words); Harriet Powell (music)

32 Six currant buns
Jean Turnbull (music)

33 I've got sixpence
Chappell Music Ltd and International Music Publications (words and music)

35 Black Friars
David Higham Associates Ltd, from *Nursery Rhymes of London Town*
by Eleanor Farjeon (words); Leon Rosselson (music)

37 Two in a boat
'From *The Best Singing Games for Children of All Ages* by Edgar S Bley
© 1957 by Sterling Publishing Co Inc, 2 Park Avenue, New York, NY'
(words and music)

41 Three old men
B T Batsford Ltd, from *The Children's Book of Comic Verse* chosen by
Christopher Logue (words); Harriet Powell (music)

43 The appleman
Epworth Press, from *A Pocketful of Rhymes* by Helen Clyde (words);
Margaret O'Shea (music)

We would like to thank Heinemann Educational Books and The Scottish Primary
Mathematics Group for allowing us to include the words of Nos 6, 7, 10, 11, 16, 21, 22,
23, 24, 32, 39 and 40; from *Infant Mathematics* First Stage, Teachers' Notes.

We would like to thank Kaye Umansky for her considerable contribution to this book – the
words and music of 34 My first week at school, 36 The centipede, 42 The King of Roly
Poly, 44 Noah's birthday – the words of 27 Nine Hairy Monsters, 38 Twenty tomatoes.

Every effort has been made to trace and acknowledge copyright owners. If any right has
been omitted, the publishers offer their apologies and will rectify this in subsequent
editions following notification.

Index of first lines